NATURAL BRIDGES

THE STORY *of* WIND, WATER & SAND

by DAVID PETERSEN

FRONT COVER: *Owachomo Bridge and pool.*
BACK COVER: *Owachomo Bridge*
FULL TITLE PAGE: *Owachomo Bridge*
OPPOSITE: *Milky Way as seen above Owachomo Bridge*

Copyright © 2014 by Canyonlands Natural History Association

Photographs copyright © 2014 by:
CHARLIE CHOC: pages 7, 11, back cover inset
WILLIE HOLDMAN: page 5
GEORGE H. H. HUEY: front cover; pages iv, 8
PAUL LANTZ: page 27 top
MARC MUENCH: pages 4, 10, 24
NATIONAL PARK SERVICE: pages 3, 6, 13 (Bill Von Allmen), 26, 29
JERRY SHUE: pages 12, 27 bottom
TOM TILL: pages i, 14, 25, 28, 30, 31, 32
WALLY PACHOLKA: pages iii, 34-35
GARY VOESTE: pages 2, 15, 16, 20-21, 23, back cover background

While reasonable care has been exercised in the preparation of this book,
the publisher and authors assume no responsibility for errors or omissions,
or for damages resulting from the use of the information contained herein.

All rights reserved. This book may not be reproduced in whole or in part, by any means
(with the exception of short quotes for the purpose of review), without permission of the
publisher. For information, address Permissions, Canyonlands Natural History Association,
3015 South Highway 191, Moab, UT, 84532.
www.cnha.org

Printed in China

FIRST EDITION 1990
SECOND EDITION 1999
THIRD EDITION 2014

ISBN 0-937407-02-X

CONTENTS

DISCOVERY
1

BRIDGE BUILDING
9

THE ANCIENT ONES
17

LIFE AT THE BRIDGES
25

UNDER A HORSE'S BELLY
33

DISCOVERY

WE MAY NEVER KNOW what Cass Hite, his two fellow prospectors, and their guide, a Paiute Indian named Joe, expected to find when they left their camp on the Colorado River one day in September of 1883, and entered the mouth of a big sandstone canyon twisting off to the southeast. Joe's people had inhabited the Colorado Plateau for more than 500 years. They knew of the natural bridges of White Canyon, and even had a name for them, a name that translates to "under a horse's belly between his fore and hind legs," an accurate enough analogy for a natural bridge's appearance when viewed from beneath.

It's probable that the Paiute guide had seen, or at least heard tell of, the sights that awaited the party up the canyon. These included three massive sandstone bridges spanning the chasms of White and its tributary canyon, Armstrong, plus dozens of remarkably well preserved prehistoric ruins perched like the nests of cliff swallows along the steep canyon walls. Still, there is no record that Cass Hite, the leader of this small expedition, had any foreknowledge of the wonders history would one day credit him with having discovered in the sandstone canyons of southeastern Utah.

Discovered may not be the best word to describe the Hite party's accomplishment, or, for that matter, the accomplishments of any modern explorers. *Rediscovered,* or, better yet, *made widely known*

LEFT: *Kachina Natural Bridge*
RIGHT: *Prospector Cass Hite "found" the bridges while exploring White Canyon in 1883. The Navajo who knew of Hite called him Hosteen Pish-La-Ki, man who works with silver.*

would be more accurate, because the three natural bridges recognized today by the Hopi names Sipapu, Kachina, and Owachomo had been around for thousands of years or so before Hite encountered them. And for hundreds of years, Native Americans had climbed on, walked across, painted, engraved the surfaces of, and even lived beneath the bridges.

At least a few early non-Native hunters and fur trappers must have passed beneath the bridges long before Cass Hite got his turn. Still, Hite was the first man to announce the bridges to the world, so he generally gets credit for their discovery. But his impromptu names for them—President, Senator, and Congressman—did not stick.

In the quarter century following Hite's description of the bridges, an almost continuous parade of survey, research, and sightseeing expeditions traveled to the canyons of Cedar Mesa, a plateau in southeastern Utah that encompasses Natural Bridges National Monument. There, they measured, photographed, mapped, and wrote about the trio of monolithic structures, the canyons they span, and the nearby "Cliff Dweller" ruins. And expeditions these visits were, given that it required several days of tough travel and camping to reach the bridges from just about anywhere.

ABOVE: *Ancient ruins near Cedar Mesa.* OPPOSITE: *Exploration party beneath Kachina Bridge.*

Still, the curious came, saw and renamed the three bridges for friends, relatives, and benefactors. Cass Hite's President was re-christened Augusta. The middle bridge became Caroline, while the third and smallest span has had at least four names in the past century, with Hite's Congressman being followed in close order by Little Bridge, Edwin, and finally and still, Owachomo.

The field notes of William B. Douglas, leader of a 1908 government survey party to the bridges, outlines the logic that led him to select the current Hopi names:

> The pictographs on the middle bridge show the symbols of certain Hopi Indian clans. The architectural design of the houses, the implements, the pottery fragments all indicated a Hopi ancestral tribe. The Hopi, whose present location is 120 miles south, have traditions that these very clans came from the north.
>
> These facts seem to clearly indicate that the original discoverers of the bridges were ancient, prehistoric ancestors of the Hopi, speaking their language and having their customs. In view of this it would be very appropriate for the bridges to bear Hopi names, as, no doubt, they originally did.

DISCOVERY

ABOVE: *Owachomo Bridge, with its slender span, appears to be the oldest in the monument.*
RIGHT: *Experience the desert silence beneath Sipapu Bridge.*

A conspicuous feature of the little bridge is a conical mount, on top of, and forming a part of the bridge rock… This feature suggests the name of "Owachomo" (Rock Mound) Bridge, a name in actual use by the Hopi.

The middle bridge is in reality named by the symbols carved on it, requiring only an interpretation. Of these is the Lightning Snake, a symbol which was painted on the bodies of the Kachina (the Sacred Dancers)…This bridge logically becomes the Kachina (the Bridge of the Sacred Dancers).

The White Canyon bridge, the largest and most impressive of the three, forms a great portal across the canyon through which all who follow the canyon trail must pass…[suggesting] the Sipapu, which according to the cosmogonic mythology of the Hopi and kindred tribes, is the gateway through which man comes to life from the underworld, and through which he must finally depart. Hence the name, the Sipapu Bridge (the Gateway of Life).

Zeke Johnson, the first custodian of Natural Bridges, beside the Goblet of Venus (no longer standing)

Although some of Douglas's bold genealogical deductions might generate some argument among contemporary anthropologists, few would disagree that these traditional Hopi names are far more appropriate than any of the previous ones.

Just months before Douglas's visit, the natural bridges of Cedar Mesa had garnered sufficient scientific, media, and public interest, bolstered by grassroots support from the citizens of the nearby settlements of Blanding and Monticello, to prompt President Theodore Roosevelt to invoke the American Antiquities Act of 1906. On April 16, 1908, by presidential proclamation, the trio of stone spans were made a national monument—the first one to be designated in Utah.

From the proclamation:

> Whereas, a number of natural bridges situated in southeastern Utah, having heights more lofty and spans far greater than any heretofore known to exist, are of the greatest scientific interest, and it appears that the public interests would be promoted by reserving these extraordinary examples of stream erosion with as much land as necessary for the proper protection thereof, [I] do hereby set aside…Natural Bridges National Monument.

A national monument was created. But the genesis of the great stone spans from which the monument takes its name had come millions of years earlier.

Kachina Bridge, the most massive and considered the youngest the of three spans.

BRIDGE BUILDING

SOME 225 TO 280 MILLION YEARS AGO, during the Permian geological period, the area now encompassing Natural Bridges National Monument was a broad, flat expanse of shoreline sands lying along the eastern edge of an extensive ancient sea. One of the most visually interesting of the geologic remnants of this time is a phenomenon known as cross-bedding. One theory holds that cross-bedded deposits at Natural Bridges were formed when dryland sands, driven by fierce desert winds, drifted into layered dunes. Each time the winds shifted, new dunes formed atop old, the lines of their layers abutting, overlapping, and interfingering. A second, and possibly more substantial theory, suggests that the majority of the cross-bedding here took place not on dry land, but when waterborne sand was deposited on submerged bars along the ancient sea's shallow shores. Over millions of years, these cross-bedded sands, whatever their origins, became compressed and chemically bonded, forming the Cedar Mesa Sandstone from which the natural bridges were cut.

As a result of later uplifting, tilting, and erosion of the Colorado Plateau (an oval-shaped, high-desert basin that centers just west of the Four Corners of Utah, Colorado, New Mexico and Arizona), thousands of cross-bedded dunes have been exposed in canyon country.

Above the thick Cedar Mesa Sandstone (as thick as 1,200 feet in places) lies a relatively thin red strata of siltstones, mudstones, and marine shale deposited when a shallow sea inundated the ancient desert. These make up the Moenkopi Formation. Although the Moenkopi contains no cross-bedding, it is nonetheless rich in rel-

OPPOSITE:
Owachomo Natural Bridge

This pool of water, close to Owachomo Bridge, is referred to as Zeke's bathtub.

ics of its own time. These include the fossilized remains and tracks of reptiles and amphibians, and intricate ripple patterns formed in low-lying tidal flats.

Atop these older formations lie younger deposits: the varicolored shales of the Chinle Formation where fossils are occasionally found; the reddish, cliff-forming sandstones of the Wingate Formation; plus several others, each in its turn deposited by sedimentation, then substantially withdrawn by erosion, all in the geologic long ago.

Some ten million years ago, the ancestral Colorado River and its tributary streams (the Escalante, Dirty Devil, Green, and San Juan Rivers) began carving into the face of these stone layers. Because the surface of the land was generally flat, these infant streams, each following its own circuitous path-of-least-resistance, flowed in meandering loops that often doubled back on themselves like ribbon candy.

About the same time that the Colorado River drainage was entrenching itself, the tremendous pressure generated by renewed continental drift slowly began warping the Cedar Mesa area, a mere hundreth of an inch a year, upward. This upwarping forced the stacked rock layers to tilt, fault, crack and slip. As the land rose and buckled, the erosive action of the rivers and their feeder streams cut downward, maintaining their original courses and elevations.

This long, slow process created the deep, winding gorges, or entrenched meanders, of White and Armstrong Canyons and their tributaries. Later on during this land-uplifiing, stream-cutting, and meander-entrenching process, the three natural bridges of White and Armstrong Canyons, as well as others that have long since collapsed, were sculpted out of the Cedar Mesa Sandstone.

Visualizing the formation of a natural bridge is made easier by thinking of the spans not as bridges, but as thin stone walls, or fins, with large holes drilled through them. This is more than an analogy because hydraulic drilling is the process by which all natural bridges are cut.

The streams that originally carved (and continue to enlarge) White and Armstrong Canyons are intermittent, flowing only when snowmelt or summer rains supply them with a torrent of water. Beginning as converging trickles in the upper reaches of the

Sipapu Bridge is the second largest natural bridge in the world.

canyons, runoff from a summer thunderstorm builds in both volume and momentum as gravity pulls it ever down. The result is a speeding freight train of water, the legendary flash flood. The lower and larger the canyon, the bigger and faster the train, and the greater the erosive freight it carries.

Rumbling along at the head of a flash flood is a boiling, churning mass of debris ranging in size from pebbles and limbs to near-boulders and mature trees. This supercargo of hydro-powered trash scrapes and cuts into the canyon walls that contain its fury. On the outside of meander bends, centrifugal force further boosts the speed and power of the flood, while inertia commands the flow to track straight ahead into the facing canyon wall.

But there is more to the cutting action of a flash flood. The rampaging stream also carries in its turgid waters tons of suspended silt, sand, and other fine grit. And this grit, while not as aggressive and fast-working an agent of erosion as hurtling stones and tree trunks, nonetheless grinds relentlessly away.

When a flash flood goes to work on a thin wall of rock that rises on the outside bend of one of its meanders, the result is rapid (on the geologic time scale) undercutting. Meanwhile, around the meander, identical forces are grinding away at a spot on the opposite side of the same wall. Aided by

Flash floods can turn a dry wash into a raging torrent in minutes

Over millions of years the erosiaonal forces of wind and water carve canyons and natural bridges.

the slower but persistent erosional forces of precipitation, percolation, and wind, the stream will eventually bore through the wall, creating a straight line shortcut that leaves the original meander high and dry. But more work remains to be done before there is a bridge.

For a while, flash floods will continue to enlarge the opening, as is still happening with Kachina, the youngest of the three Cedar Mesa bridges. But even after stream erosion has lowered the canyon floor enough to hold future floodwaters safely below the reach of the bridge's underside, other erosional forces remain on the job. Under its own weight, the span will become stressed, causing fissures to form along lines of internal weakness, occasionally setting free large slabs of stone. As always, the unstoppable trio of wind, precipitation, and percolation is going about its slow business, with surface moisture percolating down into the porous rock and dissolving the chemical cements that bind the grainy sandstone together. As this water freezes and thaws, expands and contracts, it wedges the rock apart a grain at a time, slowly converting stone to sand.

Erosion not only enlarges but also rounds and smoothes the surfaces of the opening. In time, a natural bridge emerges, but erosion doesn't stop there. It continues to enlarge the opening and thin its span. The more eroded and frail a bridge becomes, the older it is said to be. Of the three monument bridges, Owachomo, a slender and deeply fissured span, appears to be the oldest.

Although most natural bridges, including all three at the monument, have flat-topped spans, others, such as high-arched Rainbow Bridge at Lake Powell, the largest natural bridge in the world (followed closely by Sipapu and Kachina), are anything but flat. It is not, then, a span's resemblance to a manmade overpass that distinguishes it as a natural bridge, but the fact that it was cut by moving water. This requisite, in turn, suggests that most natural bridges should span canyons with streams flowing

Flashflood waterfall in Tuwa Canyon, Natural Bridges National Monument

through them. While most, in fact, do, there are exceptions, including Owachomo, whose mother stream has long since disappeared.

An arch, like a bridge, is formed when water erosion eats a hole through a relatively thin wall of stone. But with arches, the water does its sculpting work imperceptibly, from within, rather than dramatically, from without. The same erosional forces that help to smooth and shape a young natural bridge do all the work in the making of arches.

This is how it progresses: a crack or recess traps water and enlarges through erosion until it becomes a cliffside hole. More time, more erosion, and the hole becomes a cave or alcove, until water and wind eventually carve a hole completely through the fin. While it was long thought that wind was the primary sculptor of arches, it is now known, to call upon a geological aphorism, that "water does the work; wind just helps clean up the mess."

Studying an arch, or a natural bridge, you may note dark, lustrous stains streaking down the face of the wall in which the opening appears. This is desert varnish, a common sight at Natural Bridges and throughout canyon country. The main source of desert varnish is windblown soil. Clay particles in the soil adhere to moist rock surfaces. Minute amounts of iron and manganese are then attracted to the negatively charged clay particles. Over time, a thin dark veneer of minerals builds up, creating desert varnish.

In prehistoric times, these darkened stone faces served as palettes for the native rock artists of the Natural Bridges area. Their highly stylized pictographs (paintings) and petroglyphs (pecked or abraded engravings), cryptic though they are, may offer us clues as to how these mysterious ancient ones viewed their world.

Desert varnish on Sipapu Natural Bridge

THE ANCIENT ONES

LIKE THE BRIDGES THEMSELVES, the people who left haunting images on canyon walls throughout the Four Corners area have worn a variety of names. To the first Mormons who settled Utah, they were Moki. To the turn-of-the-century adventurers, who found and, all too often, plundered their shallow graves and the ruins of their homes, they were Cliff Dwellers. And for decades most of us have known them as Anasazi, a Navajo word that translates roughly to "enemy ancestors." Puebloan people today find the term objectionable. The designation "ancestral Pueblo" is now preferred, and is the term used in this text.

The Cedar Mesa Sandstone of Natural Bridges displays rock inscriptions of the ancestral Puebloans and other prehistoric peoples, which were painted onto or pecked into the rock over a span of hundreds, perhaps thousands of years. You will recall that William B. Douglas named Kachina Bridge for the artful depictions of the Hopi Kachina "Lightning Snake" he found there. Other images found at Natural Bridges rock inscription sites include eerie, humanlike figures; human handprints and footprints; animal, bird, and insect images; animal and bird tracks; and myriad geometric and abstract designs.

Painting and etching weren't the only skills the ancestral Puebloans developed. They were also stone masons, weavers and potters, skilled hunters, and innovative dryland farmers. Before their gradual disappearance from the Four Corners region late in the thirteenth century, the ancestral Puebloans had developed an intricate cosmology that we may never fully understand or appreciate. Even after more than a century of determined archaeological survey, excavation, laboratory study, and

OPPOSITE:
Ruin in White Canyon

educated speculation, the surface of the ancestral Puebloan world, quite literally, has barely been scratched.

Natural Bridges National Monument lies near the center of what was, for over a thousand years, ancestral Puebloan land. Within its boundaries, Natural Bridges National Monument contains hundreds of documented prehistoric sites. These include rock-inscription panels, campsites, dwellings, defensive walls, and underground ceremonial chambers called kivas. The Hopi use identical chambers today.

Most of the sites at Natural Bridges are modest, such as the scant remnants, situated inside the bounds of the monument's eight-mile scenic loop drive, of a small dwelling with attached granary called Loop Road Ruin. But a few such as Horsecollar Ruin down in White Canyon between Sipapu and Kachina Bridges are substantial in scope, remarkable in architectural detail, and beautifully preserved.

Native habitation of the Colorado Plateau in the Bridges area is divided by archaeologists into four primary cultural and time periods, some of which overlap. These are the Paleo-Indian (11,500 to 7,000 years ago), Archaic (7,000 to 2,000 years ago), ancestral Pueblo (roughly 2,000 to 700 years ago), and Numic speaking peoples (contemporary Utes and Paiutes).

Some evidence, in the form of stone implements, suggests that the Paleo-Indians did utilize the Natural Bridges National Monument area, perhaps as long ago as 10,000 years. However, these wanderers probably did not live at Natural Bridges. They merely passed through from time to time in the course of their seasonal hunting and gathering rounds.

The only evidence of human use during the Archaic period (7,000 to 2,000 years ago) is a leaf-shaped stone projectile point picked up by an archaeological survey team in 1961. It was during the late Archaic period, around 2,200 years ago, that low-intensity farming was introduced into the Four Corners area. The first crop was maize (corn), which not only could, with expert care, survive the semi-arid climate, but could also be dried and stored for extended periods. Human use of Natural Bridges occurred primarily during the ancestral Puebloan period (2,000 to 700 years ago). This cultural period is further divided into the Basketmaker and Pueblo Cultures. The Basketmakers, the earlier of the two ancestral Puebloan cultures, used the atlatl (a primitive but highly effective spear launcher) to hunt everything from rabbits to deer. They also gathered edible, medicinal, and ceremonial wild plants, grew maize and squash, wove elegant baskets and tough sandals from natural fibers (primarily yucca and the bark of cliffrose and Utah juniper), and lived on the mesa tops and within alcoves down in the canyons.

Basketmaker II and Basketmaker III (2,000 to 1,300 years ago) sites are well represented at Natural Bridges. During the Basketmaker III period, beans were added to the short list of cultivated crops, and wild turkeys were domesticated, their large, supple feathers used in the manufacture of warm, if somewhat prickly, clothing and blankets, their delicious meat prized as a food These Basketmakers had also discarded the atlatl in favor of the bow and arrow and were making gray or black-on-gray pottery. In addition to alcove dwellings, the Basketmaker III people lived in mesa-top pithouses.

For reasons not fully understood, only a single Pueblo I (1,300 to 1,100 years ago) site has been located at Natural Bridges. These people produced corrugated white-on-black and red orange pottery. They also erected pueblo-type masonry dwellings and storage structures atop older existing mesa-top pithouses.

The Pueblo II and III (1,000 to 700 years ago) most nearly typified what many people perceive to be the classic ancestral Puebloan lifestyle. The Pueblo II were, above all, farmers. But since both White and Armstrong Canyons are narrow and periodically scoured by flash floods, crops in the Natural Bridges area had to be planted on benches elevated above the canyon floor or, more often, up on the mesa tops.

The ancestral Pueblo lived in extended family or clan communities of a dozen or so people each, with the population of the Natural Bridges area probably never exceeding a few such groups at any given time. A typical family would set up housekeeping in small, rectangular, often interconnected pueblo rooms located beneath the protection of a south- or west-facing canyon alcove. In most cases, dwellings and storage buildings were clustered around a circular, usually subsurface kiva, with this chamber's flat roof serving as a community plaza.

The kiva was entered by way of a wooden ladder extending down through an opening in the center of the roof. There clan members, warmed and provided with the flickering light of an open fire, passed much of their leisure time talking, singing, chanting, smoking, making music on drums flutes and rasps, working at various arts and crafts. It is also believed that many of the most sacred religious ceremonies were conducted down in the smoky, cavelike kivas.

Since Natural Bridges National Monument is located at the northern extreme of southeastern Utah's primitive farming belt, many archaeologists familiar with the area feel that the ancestral Puebloans living there must have been a fairly mobile lot, moving from place to place in seasonal rounds. During the heat of the summer and

OVERLEAF: *Handprints of the Ancients*

Kelley

American Anthro

Southern Paiute shamanism
(spirit helper (shaman)
and peckings

It's not difficult to find
example ppl's engineering

and each category,
don't describe the spot
too much → jump
right into its portal

through early fall, they probably occupied their masonry pueblos down in the cool, shady, well-watered canyons. Come fall with its chilly nights, they may have moved up to the pueblos and pithouses on top of the warmer, sun-bathed mesas. Here they might have stayed for perhaps a month before drifting on south a few dozen miles, dropping down into the lower elevations and more hospitable wintertime climate of Grand Gulch. Come spring, the pattern would have been reversed, the people moving back to their mesa-top pueblos for a month or so, then returning to the canyons for the summer.

As the Pueblo III period matured, native populations throughout the Four Corners area increased dramatically, precipitating bigger communities and the construction of larger, often elaborate pueblo dwellings epitomized by the magnificent cliff-hanging villages of Mesa Verde National Park in Colorado and the sprawling pueblos of Chaco Culture National Historical Park in New Mexico. Natural Bridges claims many Pueblo III sites, all beautiful, but all quite modest in comparison to those at Mesa Verde and Chaco.

The most impressive ancestral Puebloan site at Natural Bridges is also one of the most accessible. Horsecollar Ruin is actually two adjacent ruins, distinguished as north and south. Oddly, the most aesthetic structures within the Horsecollar complex are not the dwellings, but two large, almost identical beehive-shaped granaries. It's the inverted teardrop shape of the twin granaries' doors that give Horsecollar Ruin its name.

In addition to the two large granaries, the site includes several dwellings, some smaller storage bins, and two beautifully preserved kivas. The inside of one of the Horsecollar kivas looks much like it would have 800 years ago when it was being used. There are shelves and benches recessed into the circular walls, a plastered fire pit with stone windbreak, a ventilation shaft through which the fire drew its oxygen, even a dimple in the floor a few feet behind of the fire pit suggestive of the Sipapu, the sacred gateway to the mythological lower worlds described in the cosmologies of contemporary Pueblos.

Many archaeologists feel that the decreasing number of ancestral Puebloan sites built at Natural Bridges between the Pueblo II/III interface and the mature Pueblo III period indicates that the ancient ones had already begun their withdrawal from the area. This hypothesis is substantiated by the fact that no Pueblo IV or V (750 years ago to the present) sites exist within the monument. By 750 years ago Natural Bridges had been abandoned. In the broader view, by 722 years ago the ancestral Puebloans had deserted the entire Four Corners region, moving south to resettle in the Rio Grande Valley near present-day Santa Fe, New Mexico, and on the Hopi mesas in northern

Horsecollar Ruin

Arizona. The Pueblo people living in these areas today are believed to be the descendants of the Four Corners ancestral Puebloans.

Even though we can be reasonably sure where the ancestral Puebloans went, we still aren't certain why they left their ancient homelands. It is highly probable, however, that several converging problems led to this mysterious exodus. One documented natural disaster, and a dandy it would have been for a culture almost totally dependent on farming, was a twenty-eight-year drought, the occurrence and duration of which have been scientifically determined by tree-ring-growth studies.

Additionally, after generations of relatively easy living, the late ancestral Puebloans may well have increased their population beyond the carrying capacity of their semi-arid and fragile homeland. Problems stemming from overpopulation likely would have included soil depletion resulting from overfarming, the disappearance of wild game due to overhunting, and deforestation from overcutting over timber for building materials and fuel. It is easy to imagine what further troubles the pressures generated by such severe and ongoing stresses could have spawned.

For whatever reasons, the ancient ones are gone, their tenure at Natural Bridges marked today only by fading rock inscription panels, shards of broken pottery, a few crumbling masonry walls and our ability to imagine the past.

LIFE AT THE BRIDGES

The ancestral Puebloans may be gone, but life continues at Natural Bridges…a great abundance of life. Neil M. Judd, writing of his 1907 expedition to the bridges, recalled a twilight meeting with one memorable representative of that life:

What really left a lasting impression at that one-night camp under Carolyn [Kachina] Bridge was my unexpected encounter with a mountain lion. The animal had come to drink at a trail-side pool and failed to notice my presence. However, just as I was drawing a bucket of water from a deep crevice in the canyon floor, that cougar let go with one of the blood-curdling screams for which its kind is noted…I broke the world's high-jump record right there. The lion left the scene unceremoniously and so did I.

It's not likely that you'll see a mountain lion at Natural Bridges these days. Not because they aren't there; they are, a precious few of them. But the big cat is primarily nocturnal and exceedingly shy.

Nor are you likely to spot a bobcat. Although these twenty- to thirty-pound felines do inhabit the monument, they are every bit as ghostlike in their movements as their bigger cousins.

ABOVE:
Female mountain lion
OPPOSITE:
Sipapu Natural Bridge

Kit Fox

One large predator you might just see, and an animal you stand at least a fair chance of hearing if you're out and about at sunrise or just after dark, is the durable coyote.

Far shyer than the coyote is the diminutive, almost totally nocturnal desert kit fox. Averaging only ten inches high and eighteen inches long, and weighing from four to six pounds, the kit fox is rarely seen. Even so, its rounded, catlike tracks regularly dimple the damp, sandy canyon bottoms, mute evidence of this fox's silent hunting rounds in the night.

All of these predators must eat. The western mountain lion's staple food is mule deer, of which Natural Bridges has a sufficient supply. Less often, lions will hunt desert bighorn sheep, a herd of which roams the lower stretches of White Canyon, along the shores of Lake Powell and the Colorado River. (Occasionally, the herd ventures near Natural Bridges, but they are rare within the monument.) Mountain lions also take occasional old, ill, or newborn elk, a small herd of which was transplanted in the fall of 1988 to the conifer and aspen forests flanking the Bears Ears (those twin, blunt-topped mountains jutting up to the northeast of the monument).

The bobcat's primary prey are cottontails and jackrabbits. Coyotes, too, pursue them but subsist primarily on plant foods, opportunistic carrion, and mice, voles, pocket gophers, antelope squirrels, and other rodents.

LIFE AT THE BRIDGES

These rodents are prime prey not just for various four-footed predators, but for the flying hunters as well, golden eagles, and a variety of hawks and falcons, and that most persistent and successful airborne pursuer of small nocturnal creatures, the great horned owl.

Raven

Of course, most birds that utilize the Natural Bridges area are not predators at all. A few like the talkative, jet black raven and the black-and-white magpie, are either full-time or opportunistic carrion eaters. Most smaller birds, however, subsist on insects, nuts, seeds, grains, or a combination thereof. These include the violet-green swallow, the little canyon wren with its delicate, flutelike song, described by ornithologist Roger Tory Peterson as a "gushing cadence of clear curved notes tripping down the scale," and the white-throated swift with its graceful, backswept wings and twinkling flight.

But even more numerous than the birds at Natural Bridges—in fact, probably the most numerous visible creatures through canyon country—are the reptiles. Of these, Natural Bridges has an abundance, particularly lizards, including the skittering, sleek-bodied whiptail; the side-blotched lizard with its spiny skin and blue markings behind the front legs; the eastern fence lizard with its rough hide and blue patches on throat and belly; and the beautifully marked collared lizard.

One reptile you aren't likely to see, and one most visitors in fact hope not to see, is the midget faded rattlesnake. Sightings are often reported in and around the residential area at Natural Bridges, however. Several studies were undertaken in an attempt to discover why these rattlers were so abundant there (in one year alone, twenty-five were captured and relocated), compared to other areas in the monument where they are rare. In one study, snakes were captured and anesthetized and tiny radio transmitters were implanted beneath their skins so as not to interfere with their normal activities. The comings and goings of these "wired" snakes were electronically monitored throughout the warmer months when they are most active. As far as researchers have been able to determine, the residential area straddles an intersection of ancient and popular midget faded rattler "trails," with several (at least three) dens situated nearby. The location may also be a traditional mating area. No one knows for sure.

Lizards, including the colorful collared lizard, are one of the most frequently seen animals at Natural Bridges.

Rainwater potholes in Cedar Mesa sandstone

Among the tiniest members of the animal kingdom at Natural Bridges are those that live out their brief lives in potholes. Excavated by the erosive action of standing water (which does the work) and wind (which cleans out the debris), potholes are depressions found in hard sandstone, or slickrock. When filled with rainwater or snowmelt, they provide a vital link in the canyon country's ecology.

The smallest, shallowest pools generally contain only protozoans. The sandy bottoms of slightly larger potholes may be covered with swarms of midge (gnat) larvae, which resemble tiny worms. In the biggest, deepest, longest-lasting potholes, some of which approach the size of backyard swimming pools, the patient and attentive eye can discern a variety of insects: whirligig beetles, water boatmen, water striders, backswimmers. Also likely to be present are snails and various crustaceans including minute water fleas and tiny fairy, tadpole and clam shrimps.

The largest member of the pothole gang is the remarkable spadefoot toad. This amphibian spends most of its life sleeping underground, surfacing only for occasional nocturnal insect hunts, and when its summertime breeding season and rainfall coincide. From the moment the female spadefoot deposits her eggs in a pothole, until the hatched tadpoles develop into young toads that can hop off and burrow into the safety of the sand, it's a race between growth and death by dehydration.

One trait shared by all of the pothole dwellers, from microscopic protozoans to spadefoot toads, is an incredible hardiness. All of them in their own proven way must be able to withstand summertime surface temperatures as high as 140 degrees Fahrenheit, long periods locked in the deep freeze of winter, heavy doses of ultraviolet radiation from a relentless sun, and severe dehydration when the pools dry up.

Most canyon country plants are similarly well adapted. Plant life at Natural Bridges occupies distinct communities with overlapping interfaces. Highest in altitude is the mesa-top "pygmy forest" of gnarled pinyon pine and shag-barked Utah juniper, an ecology locally referred to as the pinyon-juniper, or P-J.

Sprinkled amongst the P-J are myriad wildflowers that bloom in overlapping seasons, so that the canyon country, except in the dead of winter, is almost never at a loss for color: the oranges to lipstick reds of Indian paintbrush, the clean whites and soft yellows of alyssum, the fire-engine-red of penstemon, the varicolored wild lily, the purple aster, the lacy golden-yellow of prince's plume, and many others.

Claretcup cactus

Also at home in the P-J is the yucca, with its rigid central stalk of waxy, shell-white flowers; the leafless, joint-stalked ephedra (Mormon tea); and the bushes—squawbush, blue-green sagebrush, golden flowering rabbitbrush, four-winged saltbush, creamy flowered cliffrose. Cactus thrive in the P-J, the most common being the many-spined pricklypear with its brilliant blossoms, and the gorgeous, lavender-flowered hedgehog.

On the almost bare slopes of canyons, vegetation is less plentiful, with growing sites limited to crevices and small pockets of sand. Here grow serviceberry, mountain mahogany, Gambel (scrub) oak, and many of the same wildflowers found on the mesa tops.

Pricklypear Cactus

Less expected is the tall Douglas fir, adapted not for the canyon country but for the mountains. It survives at Natural Bridges, nonetheless, favoring relatively cool north- and east-facing slopes. The Douglas fir brings with it a community of smaller dependent plants, effectively creating a miniature montane community.

Another surprise at Bridges is the giant, rough-barked ponderosa pine. Unlike the Douglas fir, the ponderosa stands alone, distributed in patches across dry, south-facing slopes.

Hanging gardens, another seemingly out-of-place plant community, are found farther down the canyon slopes. These little oases live along the seams formed when thin bands of dense siltstones, shales, and mudstones of the Cutler Formation underlie the more porous Cedar Mesa Sandstone. As water percolates through the sandstone, it is blocked by the denser Cutler layer and forced to flow outward, creating seeps. Seeps appear on narrow ledges, on the faces of sheer cliffs, and back in alcoves. When a seep emerges near ground level and creates a pool, it is called a spring.

Dependent upon the springs and hanging gardens are several lush plant species—mosses, bracken, and maidenhair ferns, cave primrose, the rare sheathed death camas—which can only survive with constant moisture. The Kachina daisy, which needs more sunlight than other moisture-loving species, appears in shallow alcoves and on open slopes. This exceptionally rare daisy is found only in the vicinity of Natural Bridges and in southwestern Colorado.

Another distinct plant community at Natural Bridges, the streamside or riparian zone, occurs down in the moist, shady canyon bottoms. Many of the same shrubs that grow on the canyon slopes and mesa tops appear in this ecosystem, but the evergreens—pinyon, juniper, Douglas fir, and ponderosa pine—have largely been supplanted by water-loving trees such as cottonwood, willow, and box elder.

Leafless but nonetheless living and colorful plants known as lichens are spread across all the canyon country plant communities, growing almost everywhere they can find a rock to cling to. Lichens consist of two "married" low-order plants, an alga and a fungus. The excavation of many a pothole or arch begins when the acids

Box elder tree near Sipapu Natural Bridge

produced by lichens etch shallow depressions in the rock. There, water accumulates and begins its slow work of dissolution.

A final Natural Bridges plant looks even less plantlike than does lichen. It's a dark, wrinkled biological soil crust that forms on the surface of the sand. This crust, made up of a community of cyanobacteria, algae, and fungi is critical to the health of this desert ecosystem. Since the biological crust forms a thin shield over the loose sand, it serves the important purpose of reducing wind erosion, and its rough surface creates millions of tiny pockets that trap water, increasing absorption into the soil.

It takes years for a biological crust to form, or to heal itself after being seriously damaged. For that reason, it is important to avoid stepping on and destroying these "hidden gardens' when walking or hiking. While exploring Bridges, stay on established paths, where they exist, and keep on the slickrock and in the washes when hiking in the backcountry.

UNDER THE HORSE'S BELLY

THE EARLIEST VISITORS to the three natural bridges of Cedar Mesa had only one travel option. They walked. When non-natives began visiting Natural Bridges during the late 1800s and early 1900s, they may not have had to walk to get there, but they still had to work at it. Professor Byron Cummings, one of the most adventuresome of this lot, in a 1910 article for *National Geographic Magazine*, outlined the trek:

> To reach this interesting region, people from the north and west should leave the Denver and Rio Grande Railroad at Thompsons [Spring] and take the stage to Moab, a ride of 35 miles. From Moab one must travel by private conveyance to Monticello, 60 miles farther. At Monticello saddle horses and pack mules can be secured for the trip of 50 miles over Elk Ridge to the bridges.

The trip Cummings describes required five days of hard travel and camping. Today, the drive from Moab to Natural Bridges National Monument is a leisurely three hours over paved highways. At Natural Bridges, Bridge View Drive continues out to a series of dramatic overlooks. The next step, of course, is to stroll out to the designated viewpoints; only the Horsecollar Ruin overlook requires more than a few minutes to reach.

To fully appreciate the bridges and their canyons, put yourself "under a horse's belly" by following the foot trails down to the undersides of the great stone spans. Owachomo, the easiest of the bridges to reach,

OPPOSITE:
Sipapu Bridge Sipapu means "the place of emergence by which the Hopi believe their ancestors come into the world."
OVERLEAF:
Milky Way rising over Owachomo Bridge